The Elements of Online Journalism

The Elements of Online Journalism

Rey G. Rosales, Ph.D.

iUniverse, Inc.
New York Lincoln Shanghai

The Elements of Online Journalism

iUniverse books may be ordered through booksellers or by contacting:

iUniverse
2021 Pine Lake Road, Suite 100
Lincoln, NE 68512
www.iuniverse.com
1-800-Authors (1-800-288-4677)

ISBN-13: 978-0-595-39708-2 (pbk)
ISBN-13: 978-0-595-84114-1 (ebk)
ISBN-10: 0-595-39708-5 (pbk)
ISBN-10: 0-595-84114-7 (ebk)

Printed in the United States of America

This book is dedicated to Jenny and Jaden Rosales

Contents

Preface

The Web has now proven to be a self-sustaining medium, with lots of exciting possibilities; continually being redefined and reinvigorated by the millions of its users. Observers and computer experts are now saying that, after the bust in 2000, life on the Internet has entered a new phase commonly referred to as Web 2.0.

Web 2.0 implies that many of the exponential growth in online traffic will now be focused on social networking, establishing online communities, and allowing users to participate in the creation and modification of content. Where 1.0 was all about establishing an online presence and providing news, information, and entertainment for users to consume, Web 2.0 is all about engaging people in many ways, empowering them to become active participants in the communication process. Indeed, we have entered the age of the 'We' in new media.

Some examples of the so-called 'We' media or user-centered sites are MySpace.com and Facebook.com. These social networking sites allow anyone who signed up for an account to get in touch with one another, post photos, send messages, and ask others to list them as part of their network of "friends," which is a powerful vehicle in spreading information instantaneously, even making virtual unknowns become overnight celebrities.

Many established media firms with online news sites are finally catching on to this trend. Some popular news sites such as Washingtonpost.com and Nytimes.com now offer some user activity such as allowing comments on the site about a news story, comments on an article via a blog (weblog), and being able to play games and view an interactive report. Some news sites even allow users to post pictures, videos, tags, links and a host of other things.

Citizen journalism, blogging, community and user activity are the favorite buzzwords today in the online news business. Publishers and editors see the potential windfall that the web can offer and are now investing heavily into this venture. With today's newspaper circulation, readership, and profit margins slipping, media outfits have no choice but to embrace this new reality: the Web is now the most powerful of all the media. This behooves us into creating a unique brand of journalism that will cater to the demands of the new generation of media consumers. This new brand is called multimedia journalism.

How do we execute multimedia journalism online? What type of things do we have to do in order for our news site to succeed? What are the tools needed to be able to execute multimedia journalism, effectively? This book will examine and provide answers to all these searching questions. It will survey the online news landscape, discuss the elements of the online news enterprise, highlight success stories, offer directions and advise on to how to tread the challenging yet exciting waters. Please note that both online and multimedia journalism labels are used in this book interchangeably—that is—both describes the process or practice of reporting and presenting information on news sites.

1

Moving the media cheese

Where have all the readers gone? Consumers these days are simply turning away from books and newspapers and are spending more time on using other media such as television, radio and, most especially, the Internet. The decline in readership rate is especially remarkable among young adults who have come of age in a world of wireless and instantaneous global networks. They seemed to have replaced traditional print with new media habits such as online social networking, blogging, 'texting' and instant messaging. They are embracing new communications technology at a fast pace and are comfortable with doing business online.

The media industry now sits at a point where mass communication as we know it is turning upside down. The media cheese has indeed been moved. And this brings tremendous implications as to how people communicate and how media businesses stay competitive in this new environment.

Consider the statistics: There are more than half a billion people (580 million) in the world today who access the Internet on a regular basis, a third of them (30%) live in the United States. The Internet is now a $1 trillion dollar industry and rising. People spend more and more time online. Americans, for example, spend an average of 8 hours and 48 minutes per week on the Internet at home, according Nielsen NetRatings (Nielsennetratings.com, April 2006).

Media companies were quick, at least some of them, to notice the increasing popularity of the Net and wireless communications. They have adopted innovative steps to keep up with the change and to continue to become profitable by attracting younger audiences. Young audiences drive advertising and advertising is the lifeblood of a media company.

In 1994 up to the late 1990's, billions of dollars were poured into Internet investments. Company startups have experimented with different business models and modes of online content delivery. A rapid winnowing of the field then followed, where only a few businesses succeeded in the market while a great many failed to make money and have since folded.

But as soon as the economic bubble busted and the "exuberance" settled, the lessons have become clear as to what works online and what doesn't. Also, there has been a fast movement towards making software compatible to different platforms. Big software and Internet companies, such as Microsoft, Macromedia and Yahoo!, have sold applications that have seen widespread adoption among consumers and are now considered as "the standard."

Take the case of Macromedia, a multimedia company. The development and widespread adoption of Dreamweaver and Flash have made it easier for journalists to become more creative with the presentation of online content. Because these applications have become intuitive and user-friendly, with no need to memorize codes or learn programming, a large number of 'legacy' journalists (print and broadcast) are getting excited about online publishing. They are developing multimedia presentations and launching their own blogs (a running account of events, opinions, messages, posted on a site. It has the look and feel of a personal journal).

More online news sites out there are moving away from the habit of just posting text and photos. They do more multimedia content such as interactive maps, videos, animation and games. Final Cut Pro and Flash Professional have made it easier for journalists to add video, voice, and music to a news article.

New concepts have also emerged as a result of innovations in online journalism. Among them is the idea of citizen or participatory journalism, where visitors to a news site are allowed to contribute stories and other materials. 'Experience journalism' is also a new idea where, instead of just posting a story, an editor can also upload an interactive game feature, which for example, makes it possible for a visitor to experience an event say balancing a state budget, or going through security check at the airport, or fighting a war on the streets of Najaf, Iraq.

This book explores the new world of online journalism. It guides the reader as to how to create innovative multimedia stories. It explains the nature of today's media consumer and talks about ways to gain new users as well as sustain a high rate of return visits. The book describes the step-by-step process of producing a multimedia story and talks about other important factors of online journalism such as audience, design, promotion, ethics, job prospects, and future directions for online news.

2

Online Journalism: Elements of a digital story

The practice of online journalism is much like cooking on Emeril Lagasse's show—one must always kick it up a notch. News sites carrying a conundrum of plain text, headlines and some pictures are now passé. It was version 1.0 of doing online journalism.

Mimicking the content and design of a newspaper or magazine can no longer qualify as online journalism. This new way of reporting only kicks in when a reporter or web producer comes up with a unique way of storytelling that often maximizes the non-linear and interactive potentials of the web, giving way to a more increased user experience and participation to the story being told.

A non-linear story means the site visitor has control over the sequence of what he or she wants to see and hear. Interactive means the visitor can do something with the story, which could be in the form of playing a game or exploring an animation.

Today's practice of online journalism (version 7.0) is a lot more exciting than its predecessor. It shows less clutter, simple design and navigation, interesting content, high visual appeal and interactivity, and lot more options for the visitor. These things comprise the very ingredients of the new media reporting. It is digital storytelling using various techniques learned from radio, television, newspaper, and then putting in the interactive and other unique advantages of the Internet into the mix.

What are the elements of a multimedia story?

There are two levels of a multimedia story: Basic and advanced. The basic level includes a headline, text, picture, graphic, and related links. The more advanced level carries the following added features: audio, video, slide shows, animation, interactive features, and interactive games

Headline—Carries the title of the story that when clicked often takes a reader to a separate page for a more detailed presentation.

Text—The body of the story contained in one page or broken up into several connected links.

Picture—Powerful photo that often accompanies a story.

Graphic—This could be the logo, drawing or illustration things related to a news story.

Related links—Found either highlighted within the paragraphs of a story or to the side or bottom of a page. This is a unique feature of the web that allows a reader to have a deeper context or understanding of the background of the story.

Audio—Sound, music or voice over recorded as a stand alone or mixed with a slideshow or video.

Video—A sound bite or a separate news video package that accompanies the text of the story.

Slide Shows—A collection of photos that looks more a like a picture gallery with captions. Some slideshows have sound and voice.

Animation—Moving images produced to add more impact to the story.

Interactive Features—Graphics designed to interact with the user. Examples include timelines, maps, and other interactive files designed for the user to 'experience' the news. An example of this is the award-winning interactive site called Chicagocrime.com. The site combines the functionality of Google maps with crime statistics in and around Chicago. So one is able to pinpoint a place and get the crime rate of that particular neighborhood.

Interactive Games—Designed like a mini-video game where the visitor can "play the news." For example, a report on the Republican convention may be accompanied by a an interactive game called "Find Your Inner Republican," complete with questions as to whether or not a visitor fits into the Republican view of things.

What qualifies as a good multimedia story?

a. Stories with good quotes.
b. Stories involving processes (how to).
c. Stories involving historical understanding.
d. Stories that are simply fun for an interactive feature.
e. Complex stories, which are hard to explain via text alone.
f. Stories that might offer an interactive experience for the visitor.

3

Producing a multimedia package

In order to produce a multimedia project for an online new site, the first thing to do is to become familiar with the following applications: Photoshop, Macromedia Flash and Final Cut Pro or Avid (video editing software). Plus, it will not hurt if one knows how to use Dreamweaver, a Macromedia web design software, and perhaps Adobe Illustrator, which is used in editing photos and creating logos and graphics.

No time to learn all of them? Try at least Flash and one of the video editing applications. For starters, there is a free, strip-down version of Avid video editor, along with an easy to learn tutorial, available for downloading at avid.com. Don't be scared to try out these things because, for the most part, they are user-friendly and easy to learn.

Dreamweaver, for instance, works much like Microsoft Word. Students or journalists, with no prior experience with the software, only need to type, draw, insert text or objects, save it as an HTML file (hypertext markup language) and then they are good to go. Learn a few more advanced tricks such a using frames, flash buttons, tables, links, etc., they'll be able to master it very easily. It's fun to play with the software because one doesn't have to learn about HTML coding anymore, instead, he or she just have to concentrate on making the web page look great.

When producing a multimedia package, don't worry too much about the technical stuff or whether special effects and animations are used in the presentation. Concentrate on the storytelling, the relevance, and flow of the story being told.

The following are suggested projects journalism students or beginning online journalists should work on to master the skill of producing innovative multimedia presentations. The first project is titled as "the movie in my mind," followed by "pictures paint a thousand words," "shoutouts to my homies," then finally to a more advanced "video on demand" online news package.

1. "The movie in my mind"

   Instruction: Produce a three to five-minute digital story about a fascinating thing that happened to you or your friends, family or neighbor. It should be an idea or topic that the viewing public might be curios to know. This project is an exercise in digital storytelling using Quicktime and DVD formats. Materials needed: Digital video camera and a blank DVD disc (DVD-R)

   The goal here is for a beginning online producer to get used to video editing and storytelling using production devices such as sound, music and voice-over. This project can be done using Avid, or preferably, iMovie (for Macs) because it is so easy to use.

   Here's what to do with capturing video images from the camera to iMovie. Click on the capture button [see figure] and then click on the clips that you want to include in your video presentation. Each square on the upper right side of your screen represents a clip or a scene.

   Drag any one or all of these clips down to the timeline area and then hit play. Now you have a video package. If you add sound and music to it by dragging files to the timeline area, then you really have completed your digital story telling project in no time.

   Save your final project and then try to view it using Quicktime player. Your video is now ready to be exported to an online news website or to a recordable DVD-R. If you want to save the movie to a permanent disc, just drag the saved file to a blank DVD and thin hit the iDVD burn button. This then completes your "movie in my mind" multimedia project.

   Sample the Web. A great example for this video storytelling is IBM's Webisodes (http://www.wwpl.net/ogilvy/ibm/fusion.html). Each video episode presents a way to integrate computing technology into the different aspects of business.

2. "Pictures paint a thousand words"

   Instruction: This is a slide presentation (storytelling) complete with voice and music about an event, situation, place or persons on campus and around your local community. Materials needed: Still camera, Music CD and Flash.

The easiest way to do this is to open Flash MX 2004 Professional and just click on the photo slideshow template. Pictures will then appear on the "stage" or center screen complete with a navigation bar and a caption space.

Look into the timeline area on top and click on the frame that has the existing photo, cut the picture out (go to 'edit' on the file menu and then hit 'cut') and then replace it with your own picture (go to 'file,' click on 'import' and select 'import to stage'). Test your work by clicking on 'control' on the file menu and then hit 'play.'

As soon as you're satisfied and done with the project, save the document as a Flash file. You'll need it in case changes are to be made later on. After saving the Flash file, it is now ready to be exported as a movie (click on 'file' on the menu bar, then hit 'export' and 'export to movie'), which can be inserted to a web page for viewing on the Internet. According to Macromedia, creator of Flash, 97% of Internet users are able to view its movie file on their computer. This is due to the ready availability of a free downloadable Flash player.

If you want a more original and advanced slideshow presentation, it is a good idea to go through the tutorial first. Knowing how to execute commands such as "import to library, insert keyframe, create motion tween, create motion guide, modify background," and others is key to becoming a highly skilled Flash media producer.

Sample the Web. USA Today, MSNBC, and The New York Times Online provide excellent examples of slide shows that greatly enhance a news story.

3. "Shoutouts to my homies"

Instruction: Ask students to talk about important people in their lives on camera. Cut the dramatic clips into 10 to 15 seconds sound bites and post them to a newspaper site. Professional journalists can do the same perhaps by interviewing locals and then posting the interesting sound bites on a news site. Materials needed: Flash software, a digital video camera and Final Cut Pro or Avid video editing software.

Using Final Cut Pro or Avid (there is a free version online), cut the video footage into sound bites. Then save the files in one folder because these will be exported to Flash.

Open Flash and click on create a new document or go to File <New <New Flash document. You're now ready to import the movie to the Flash library

(File <Import <Import to Library). If you don't see the library bar to the right side of the screen, you can make it visible by clicking on Window and then checking off Library.

As soon as you see the video file(s) in the library, drag and drop any clip to the main stage—the big square work area. You should see your movie there. Test it by clicking Control <Test Movie on the top menu.

If you like what you saw, save it as Flash file (click File <Save) and then export the movie with the .swf extension (File <Export <Export Movie). You are now ready to insert this to a web page.

Note that text and other background designs can be incorporated into a Flash movie. Just go to the tool bar (Window <Toolbar) and then add text by clicking "A" in the toolbar and then start typing in the stage area. The background can also be modified (click on Window <Properties) using different colors in the properties area.

Sample the Web. Small-town online newspapers have greatly enhanced the way they serve their readers by providing them with an option to send video greetings to the website. Typically, an editor goes through the videos and might decide to post some of them on the site. Another innovation is to allow site visitors to send in video "obits" (obituary) for those who want to honor a dear departed.

4. "Video on demand"—

Instruction: Produce a one or two-minute video news or feature package for a news website. Follow the format of a video news package for local TV station. *Materials needed: Digital video camera, Final Cut Pro or Avid and Flash.*

This video on demand project is similar to a TV new package with its mode of delivery being the only difference. Which means that to create this project, all the sound bites, video clips, voiceovers, and background music should be first assembled and edited using Final Cut Pro or Avid—both are non-linear editing software.

As soon as this is done, save the file or export it to Flash. Open the Flash software, create a new document, and then import the video to the library (go to File <Import to Library).

Drag the video icon from the library (if unable to see the Library, click on Window then select Library) to the center stage, which is the rectangular work space. Test the movie file by clicking on Control <Test Movie on the top menu bar.

If satisfied with the output, save the file and then export it as Flash movie by clicking file Export <Export Movie on the menu bar. The new Flash movie can be linked as a separate file from the headline or lead of a news story or it can be put in the same page as the text of the news story.

Scripts and Story Boards. Be it a video for the web, a short animation or interactive timeline project for a news website, a successful execution of a project starts with good planning and well-crafted scripts and story boards. It cannot be emphasized enough that in order for a multimedia project to become successful, a detailed script or story boards should be made available for everyone involve in the process. This gives the team a sense of the vision of the project and the end-result that they are trying to reach.

Interactive features, animations and online games. The four projects discussed in this chapter are the most commonly found multimedia features on news sites. Other interactive features such as animations and online video games are not used frequently because they are difficult to produce, not to mention expensive and time-consuming. But these types of materials might be gaining popularity soon.

Sample the Web. MSNBC.com offers excellent examples of video news packages on the web. The site also shows interesting interactive features and games that students and online journalists alike can use as models for multimedia production.

4

Web Design for Online News

Howard Raines, the former executive editor of the New York Times, said in an Atlantic Monthly article that a newspaper in order to become successful must have a creative design, socially-interesting articles, and a culture of achievement in its newsroom. The same holds true with online news. However, this chapter focuses more on the first requirement—that is—the design of the web pages. The chapter highlights and addresses the challenges of online news design. It provides the readers with principles that will help create a more pleasant, attractive, and easy to navigate site.

10 Principles of Online News Design

1. **No bling-blings.** Stay away from animations, graphics, video and sound that do not contribute to the story or have nothing to do with the news story. All you end up doing is creating clutter and slowing down the loading process. There has to be a compelling justification for every element that you put on your news page.

   Use Metaphors. Some news sites have used metaphors as a guide to the overall look and feel of their news page and multimedia content. For example, if your newspaper has an ongoing series about a criminal investigation, perhaps the graphic designer can place a picture of a filing cabinet on the center of a page and then the drawers are active links, which take the visitor to several aspects of the investigation such as witness testimonies, photos and other evidence, police and prosecution, etc.

   The Winston-Salem Journal offers a good example on how to use a metaphor in telling a story about a criminal case on its online news site.

2. **Easy Search and Navigation**. There should not be a need for global positioning satellite just to go through your news pages. Always simplify and

don't let your visitors figure out your site. Free up the main page of your news site and limit the stories and headlines there, perhaps only 3 to 5 headlines with lead paragraphs and the rest should contain only a menu of headlines that are linked to another page. The nytimes.com website offers an example on how to do this effectively.

Use clickable thumbnails for your pictures and videos on the main and other pages.

3. **A way home.** You should always provide a consistent menu either on the left or bottom part of the news page so visitors are able to get back to the main page or to other pages of the site. People who go online do not always follow the things that a site intended them to do. For example, when visitors use a search engine like Google, they often come up with a page that is deep within your site. In other words, they don't always go through the main page to look for the document.

4. **Limit fonts and colors.** So as not to confuse visitors, limit the number of fonts and colors used on the news site. Limit font styles, types and colors from 2 to 4. This way you'll achieve consistency and the whole thing looks pleasing to the eyes.

   There should always be a good contrast between background color and text. Text and photos should standout from the background. Feel free to experiment with colors, just bear in mind that any color used conveys certain meanings. When in doubt, always use the default background color—white. Blue has also become one of the most common background colors, at least the navigation menu bars. Some news sites that feature a gallery like appeal use black as background. Again, feel free to play around and discover whatever works for your and your target readers. Just make sure that you follow the golden rule in web design—simplify, simplify, simplify.

5. **Linking and grouping.** Related to the simplify mantra, avoid clutter and putting too many links on a page. Group links together into 4 or 5 general categories that will appear on a menu bar. Do hierarchical linking where a visitor is able to see sub-categories or sub-links as soon as the cursor is placed on any of the general categories.

6. **Blogs, chats, and message boards**. Maximize the unlimited space and the capacity for instant feedback that the web affords. A news site will not be

complete without message boards where readers can react, comment or offer feedback to a certain content or topic.

Another cool thing about the Net is the ability of people to chat with authors in real time. Online producers should exploit this unique feature to drive up more readers and repeat visitors to a news site.

Some news sites have also started linking blogs (short for web logs) of reporters at the end of a story. Blogs normally contain ideas or thoughts from the writer about a news story and transcripts of exchanges between the author and the readers.

Some updates on these and other related components:

RSS Feed

There are so many cool free features out there (open source) that an online news designer can use to add value to a news site. One of them is what techies call RSS feed (Really Simple Syndication). This simply means that, by simply copying the XML code of another news site to your site, you can subscribe to the headlines of the said news site. A running list of headlines will appear on your site, whether they are from Yahoo! News or Associated Press or the New York Times.

Blogging

Wordpress is an example of an open-source application that can be used to provide a blogging feature on a site. This enables the user to post comments (blogs), which can be published alongside the news story.

Search and Maps

News sites can also utilize existing search engines, weather, and maps by adding them to the site. Even detailed statistics of traffic to a site can be obtained now with the many free software available such as Google Analytics.

Photo gallery and community boards

Being able to form community or groups within a site is now an important part in citizen journalism. This feature should not be overlooked and must be a priority in the operation of a news site. The same applies with the option of being able to upload photos on the site.

7. **Breathe, stretch, shake—let it go.** Never let your reader hold their breath too long by letting them scroll to no end. Divide a long news story, for

example, into several pages with active links at the bottom indicating the page number that the reader is on.

It's also a good idea to provide an option to print or send the article to someone via email. The email and printer-friendly options provide a great way to increase satisfaction among readers and will provide a way to drive up traffic to your site because the people who received the emailed article will more than likely click on the link to read the news story.

8. **Unobtrusive advertising.** Ads should not take over the content of your page. Banner ads should be placed either on top, bottom, or the right side of the screen. They should not interfere with the way people navigate the page or get to the content. If for example, a reader clicked on a news story and a full page animated ad pops up, there should be a way for that visitor to skip the ad instead of waiting for it to finish loading before being able to see the article.

9. **Make them log on.** This is a must-have option on a news site—that is—if it ever dreams to become a commercial success. Users who log on offer a wide variety of data as to what contents they looked at and how long they viewed them. Site managers are also able to find out about navigation patterns, content preferences, and many other insights that can help them target the right ads and improve content and design.

   User Registration

   The success of any citizen participation online is primarily based on trusting the user. It is important that a user registration must be filled out before anyone is allowed to take part in the creation of online news content. The mission statement of the news site as well as the terms of agreement should be available and be part of the rules that the user must agree to before being granted access to the site. User registration allows the online news moderator to get a clear profile of users and their patterns of navigation. It will also allow the moderator to expel any person who violated the terms of the agreement.

10. **Always test pages.** Never launch a news page without testing it with your target audience to see if it works. Programmers and site designers often

assume that users are able to figure out the site and that they share the same ideas as to what is important in a news page.

An effective news website should never ask visitors to perform certain asks before they're able to view content. In other words, site navigation should always be intuitive and functional. It should not demand a certain level of skill and practice just so people to be able to access its content.

5

Bloggers and Citizen Journalists

It is amazing how in a decade's time the Internet took off and turned journalism as we know it upside down. Today, anyone can become a media mogul even without a press office and an army of employees. With the aid of a computer and an Internet access, anybody can become an online newspaper publisher even if he or she only works from a basement in Waukegan, IL. Add luck, hard work, and creativity to the mix, the chance of being the next Rupert Murdoch is within reach.

There are two unique features which the Internet offers and the print media do not: citizen participation and dialogue in real-time. Managers of online news sites may well exploit this to a maximum in order to increase visitors and advertising revenues, and reduce labor cost.

1. **Citizen Journalists**. The Net has a bottomless news space and so there is always room for everybody to provide content. This is where the concept of citizen participation should be used. People in local communities usually have the best nose for news. They know what's going on in their neighborhood and it is only fitting that they be tapped as reporters.

   Let's take the example OhMyNews.com, a popular Korean news site where readers are allowed to post stories about their community. Editors of the site receive and evaluate stories submitted by citizen journalists. If a news story is credible, without legal and ethical problems, it is immediately posted on the site. There is an evaluation system put in place to weed out the reliable from the unreliable citizen journalists.

   Gothamgazette.com is a New York news site that also uses that help of citizen journalists. Readers from the different boroughs of New York are asked to submit tips, stories, and commentaries, which the editors read, edit, and then post those that are news worthy.

Some news sites have enriched their content by asking readers to submit audio, video and other interactive files to the website. Some sites are now beginning to post voice commentaries, obituaries on compressed video, greetings, and other digital video clips of family and community events.

There might come a time when citizen journalists are paid for their stories just like freelance writers. Most online news sites have yet to develop a system to compensate contributors. But this won't happen unless news sites begin to turn in profits for their operation.

2. **Bloggers.** This is another component of citizen journalism. Blogging, short for web logging, is now rapidly gaining in popularity. Many host sites offer a space free to anyone who has the time and the energy to maintain a blog. One of the most popular hosts for blogs is blogger.com.

   Bloggers post their own running comments or reporting of just about any topic be it politics, the economy, entertainment or social gossip. An example of a blog is savethephilippines.blogspot.com. The site owner posts comments on political, cultural and social issues relating to Philippines society.

   Aside from commentary and original reporting, blogs also provide links to news pages, documents, interactive media, pictures and graphics that are related to the blog's main focus.

   Blogging is time consuming and can easily take away limited work hours from columnists and reporters. Some editors have asked their reporters to discontinue this practice because it reduces the amount time focusing on producing a quality newspaper.

   Some independent bloggers such Matt Drudge at mattdrudge.com and others have managed to break certain stories that were picked up by the mainstream media. These bloggers have acquired a certain level of influence by setting the agenda of what the big media, especially cable television, should cover.

   For instance, during the 1st of three presidential debates of 2004, bloggers have posted a picture of president with a bulge on his back. The bloggers speculated that it was a wireless transponder and Bush was receiving instructions from his deputies backstage. This story and picture spread like wildfire on the web and was soon picked up and written about by the New York Times and Washington Post.

The power given to the people who run blogs became eminent during the 2004 Republican and Democratic conventions. For the first time in history, citizen journalists were given slates to cover the conventions.

6

Communities and Networks

After the bubble burst during the first year or two of the 21st century, investors are back with excitement about the new possibilities online. The first wave of the Net innovation was dubbed Web 1.0, which was all about racing to establish an online presence. It was all about providing information and entertainment for the rising number of online users. Entrepreneurs poured in billions of dollars for anything dotcom. Hundreds of these e-commerce sites became household names—Google and eBay—and have experienced high financial windfalls. But along with these successes came thousands of investments gone bust.

Enter the new phase of the Internet craze called Web 2.0. This is the phase showcasing the power of the individual users, networks and online communities. This phenomenon is making businesses take notice because networks have become such a hotbed of fame and wealth.

Let's take the case of Myspace.com and Facebook.com. These are all social networking sites. The former is open to all while the latter is limited to colleges and universities. Anyone who signs up for either one of them can establish his or her own page—profiles, music, photos, videos and all. The personal site allows virtually anyone to post messages and to link and add another user as a friend. It is possible for someone to theoretically create a network of millions of friends through the linking process offered by Myspace and Facebook.

Facebook.com has been very popular with college students. It's a lot tamer than Myspace, because the site has stricter policy on indecency. Facebook allows a registered user to list other users as friends. It allows message posting either on the public message wall or the private message system. It allows a user to upload photos and other files and also for so-called friends to "poke" one another. Harvard sophomore and now dropout Mark Zuckerberg founded Facebook. His original intent was to provide a means for his friends to keep in touch electronically, but then his site caught on and became a hit on campuses throughout the United States.

Myspace, on the other hand, was developed by two guys from Silicon Valley—Tom Anderson and Chris DeWolfe. It is a lot more popular than Facebook because anyone can use it. It has now become a popular means for indie bands to promote their own records. On the flipside, the site has also become a hotbed of activity for child predators who want to take advantage of their access to information posted by minors. This new openness or ready access to the site is making a lot of parents and authorities worry about privacy, indecency, and other ethical concerns. Young people who use the site tend to be very open about their life. They reveal even the most intimate details of their lives. Some are unmindful of the fact that negative information could be used against them, especially if they apply for a job or apply for admission to a college or university.

Social networking sites such as Myspace have become the object of investors' attention and affection. Rupert Murdoch bought Myspace for $580 million dollars. A Canadian couple (Stewart Butterfield and Caterina Fake) who owned the popular photo sharing site called Flickr became millionaires when Yahoo! decided to buy the site for $35 million dollars (Newsweek, April 3, 2006).

Aside for networking sites, the new deal now is about collaboration and trust. There is currently an explosion of the open source movement, where software firms put their codes online for others to use and modify, resulting in a rapid improvement of software innovation. For example, Google and Yahoo! have opened up significant aspects of their codes for all to see. An even more amazing thing online is with the aspect of motivation—that is—if people sense there is a "common good" purpose to a site or software, people are willing to spend an inordinate amount of their time improving an application.

When it comes to doing things for the common good, Craig Nemark is the champion of free online collaboration and transaction. His website, Craigslist.com, allows people to post classified ads, personal ads and announcements for free. Much to the chagrin and anger of newspaper publishers, millions of people use Craigslist because they are able to connect, sell or exchange goods and services—for free. Craigslist is causing a lot of concern in the newspaper and real estate business because people are shunning print ads and real estate agents. The result is a tremendous drop in revenues for both of these industries.

Aside from collaboration and networking, trust has become the main driver of online activities. Sites such as eBay, Amazon, Craigslist and others base their operation mostly on personal trust. For instance, eBay users exchange goods based on the premise that each one will deliver the said product as promised. Users are free to post comments and rate each bidder and seller's performance. Which then leads to a hierarchy of credible bidders. The higher you are rated and the more frequent you do business on eBay, the higher the likelihood that you will be given first priority to win the bid.

Amazon.com and Craigslist also operate based on trust. Buyers on Amazon are able to post book reviews and rate buyers and sellers on the site. The same thing applies with Craigslist. People are able to police themselves and alert the site moderator if there are any inappropriate or inaccurate postings, which will then be removed from the site.

The idea of trust and collaboration online will not be complete without mentioning the sudden rise in popularity of Wikepedia. This site is essentially a user-driven encyclopedia. It allows anyone to type an entry into the site on virtually any topic. Users are given the opportunity to alert the moderator for any inaccurate postings on the database. This site has become a popular resource for people looking for quick answers to virtually any question. This idea of working together to build an online knowledge database is not without its share of flukes. For example, a citizen from Tennessee posted inaccurate information about John Seigenthaler, the director of the Freedom Forum and former lawyer of Robert Kennedy. Mr. Seigenthaler was portrayed in the Wikepedia entry as one of the suspects investigated by authorities for the assassination of President John F. Kennedy. The truth is he was never a suspect and there was never an investigation of that sort involving him. The entry has since been deleted but not without much controversy and uproar about this open knowledge system.

The power of the individual

The individual has been bestowed with such great powers online to make a difference and to generate a movement or start a chain reaction or diffusion of information at instantaneous speed. Organizers to the biggest pro-immigration rallies in the U.S., for example, used blogs, mass emails and discussion groups to ask people to join in the nationwide rallies calling for immigrant rights, guest workers program, and paths to citizenship for the 11 million illegal immigrants living in the United States.

Another example of user power is the website YouTube.com, which allows a registered user to post videos or video clips of any kind and for everyone to view clips of interviews, shows or performances that they might have missed on television. Bands all over the world use this site to post music videos and to promote their own CD's. "The Chronicles of Narnia" comic skit by the Saturday Night Live crew became an instant hit online once it got posted on YouTube. Millions of people started sending the file to other people creating a massive traffic to the site.

Another power conferred on the individual is the ability to quickly react to stories from the mainstream media through discussion boards or blogs (short for web logs). Some people like Andrew Sullivan, a political analyst, became very popular due to their daily blogs about a specific topic, which in the case of Sullivan, mostly witty political commentary and observations.

Readers can easily alert editors of online news sites to reporting inaccuracies. They even help reporters do a better job of covering a story by offering helpful tips. Online citizens can question the accuracy of media coverage through blogs or through citizen media sites such as smokinggun.com. This investigative site blew the lid off James Frey's lies in his bestselling autobiography titled "A Million Little Pieces." The book was an Oprah book club pick, which caused the talk show queen so much anguish and embarrassment.

Collaboration and Trust: Implications to Online Journalism

What are the implications of the popularity of social networking, collaboration, and citizen participation to the practice of online journalism?

First, we are seeing the demise of the role of journalists as sole arbiter and producer of news content. The creation of news is now a two way process; feedback can be obtained even before the story gets published. This is a boon to reporters because feedback enables them to do a better job of covering a story.

Second, the old gatekeeping model is now obsolete, and so goes the high sense of self-importance for those who work in the newsroom. Online users have a lot more power now to dictate what they want to see published. This is evident in the prominence given by the *New York Times* to a section on the most emailed or most popular news stories. The instantaneous reader feedback gives the editors a cue as to what should be given priority in coverage.

Lastly, in order for online journalism to flourish, we should embrace blogging, social networking and citizen participation. In other words, journalists must actively engage the online users so there is always a reason for them to go back to the news site. The more traffic to the site the higher the ad revenues.

News sites should allow users to post their own comments, pictures, video, and audio. Online citizens want to be an active player in the creation of news content. This need for users to participate and form community has now given birth to a new form of journalistic enterprise called citizen journalism.

Citizen journalism does not imply that we no longer need the services of a professional journalist. Professional journalists are the major league players in news-gathering business. They know the ins and outs of covering and producing a credible public affairs report. The citizen journalists perhaps can be considered as minor league, but they are players nonetheless. And the whole thing makes the online journalism enterprise a lot more competitive and exciting for everybody.

7

Online News and Games

There has been so much talk of late about the decline in readership among young people. If given the choice between playing a video game and reading a news story, young people would certainly pick an Xbox rather than a newspaper. But what if we could use some video games to get them interested in news and public affairs?

Professor Ian Bogost of the Georgia Institute of Technology believes that video games can be used in getting people interested in news and in the discussion of social issues. As a video game designer, he is experimenting with games that can make people experience the news or get immersed into a highly debated issue.

Bogost designed video games that focus on news and social issues. One of his projects is takebackillinois.com, a website offering an interactive game where visitors can play and make policy decisions for themselves such as balancing the budget and making sure that governmental resources are allocated properly.

Most of the games out there, however, are still superficial in content and design. Games are mainly for entertainment, but given creativity and hard work, online news can take this to a higher level where people are not only entertained but also enlightened about issues. For the most part, games provide simulations and experiences that a newspaper story can only describe. Games provide a whole new dimension as to how media consumers view news and public affairs. Games turn media consumers into actors. They also tend to motivate players to know more and master about their content.

Some news websites have started to pepper their content with some simple and elaborate interactive games. But still most of the interactive features lack depth. They're simply there to provide something fun while a visitor is on a news site. Take the news site gothamgazette.com. It is a nice site based in New York, which relies on citizen volunteers for its news and information. It also carries a good amount of games such as balancing the New York city budget and test your inner Republican. The games are simple and straightforward, but they provide a fun way of knowing politics or public policy.

Professor Bogost runs a website called watercoolergames.org, where it features an array of examples of games raining from politics, news, events and others. Bogost asserts that most of the games still need to take the second step—that is—to have a little bit more complexity to the plot, or to provide emotion and psychological dimension to its characters.

Indeed, a majority of the interactive features and games online today are simply designed for fun and entertainment. Online news sites still have to fully exploit the potentials of interactive games by coming up with features that are not only fun but also provide educational value and allow users to think critically and to master knowledge of the issues.

The only major setback for online news games to flourish is that the games industry has not shown any interest towards the serious uses of games. Leaders in the news business see games as a threat or simply unaware about how they work.

Yet despite the tepid support from the news and gaming industries alike, there are a number of games now available that offer a good start for online news. Among them is hamstersforkerry.com, a fun way to learn about poll numbers and at the same time trying to save a hamster from drowning.

News and Games have unique characteristics that can complement each other.

While news tells a story in narrative form, games show the media consumer how to do or how to navigate (procedure). While news provides information and description of events, games offer experience, action and rules. Finally, while news contains static text and photos, games are dynamic.

If news and games were to be merged in an online environment, the byproduct of this is more powerful messages and experiences given to the media consumer. As with anything that comes with innovations, this new hybrid news environment will certainly redefine journalism as we know it. For example, perhaps the concept of journalistic balance and objectivity will come to mean as putting more options and complexities in characters in order to avoid charges of bias in news-related games.

Assignment:

1. Bring a video game to class or show the class your favorite online game and list the important characteristics as to why you like the game.

2. Name at least five news stories or recent headlines that might be suitable for an informative and fun interactive game on your news site.

8

Managing Interactive Projects

Let's say the news website has already been up and running for quite some time and that the right people—online editor, graphic designer, media producer and programmer—are well in place to make sure that the operation is running smoothly. Let's assume that the staff has already mastered the dance of updating the site, of covering breaking news, or uploading photos and other materials at a fast pace. The next thing to do now is to experiment with a higher level of digital storytelling—that is—working interactive projects.

Interactive projects provide an opportunity for the staff to show their creative juices, experiment with unique and different ways of telling a news story, which if done right, can offer a spectacular experience and can be a source of attracting new and repeat visitors to the news site.

What qualifies for an elaborate interactive project?

(1) Any news or topic that can be presented as an in-depth story (narrative), using audio clips, video, text, photos and animation. A good example of this is the US military siege of Fallujah, Iraq. The interactive team could use audio clips, video footage, reporter accounts, interviews, and interactive maps, which show how the battle started and how it ended. Casualties, costs and other important numbers may also be included in the overall packaging.

(2) Any topic that can be packaged as an easy-to-navigate information source or reference for the reader. The history of the civil rights movement in the United States is a good topic for an interactive media project because producers are able to present timelines, maps, graphics and lists, creatively.

The job of creating interactive projects gets easier when the news site is tied to another media entity such as a TV station or newspaper. The timeframe is shortened because of the ready availability of media materials.

Putting It All Together

After deciding on a topic appropriate for an interactive presentation, it's time get to work through the different stages of its production.

Stage 1: Organizing the team—You would need someone who knows how to handle content (reporter/editor), a person who is familiar with sound and video (RTV producer), somebody who knows how to make good graphics (graphic designer), and you need a person who knows the technology of the web (coder or programmer).

Stage 2: Planning—The stage where the team starts to brainstorm ideas to be included in the package. A comprehensive calendar of activities is drawn, broken down into details such as the scope of the project, assignments, deadlines and expected outcomes.

Stage 3: Data gathering—All the people involved with the project go out to do research, collect relevant data about the story. If necessary, the team must go out and conduct field research and interview primary sources.

Stage 4: Production and coding—Everyone does his or her assigned task from editing text, to cropping photos and editing video clips. The team often makes adjustments to the plan due to content or technical limitations. When the deadline comes, the team is expected to deliver on the assembled interactive project for testing.

Stage 5: Test and revise—The entire online staff, or even outside personnel, should be invited to review the entire project. Look for typos, grammatical errors, technical bugs, design flaw and other errors that will negatively impact the overall experience of the site visitor. The review process should be done over and over again, checking for consistency in navigation and appearance. If flaws are found, the person in charge should readily fix the problem and the whole team should constantly refine the presentation until it has achieved a degree of clarity and coherence.

Stage 6: Going Live—At this stage, the senior editors should have already been given ample time to do a second review of the material. A decision should be made the level of prominence that the interactive project will have on the site. Is it going to be on the main page? Is it going to be part of the Multimedia link?

The interactive project should also be heavily promoted on other media—newspapers and television—to boost the number of visitors to the site.

Stage 7: Post-launching—This is the time to do an evaluation of the project, talk about its effectiveness and to see how well it was executed. The following questions should be asked: are there any aspects of the project that need to be tweaked? Was there any feedback from readers? Did the project generate additional traffic to the site? What were the insights or lessoned learned from the experience that might be useful in improving future interactive production work?

9

Internet Security and Ethical Challenges

The Internet is a powerful medium for instantaneous global communication and information sharing. It is also an efficient tool for business and commerce. This medium, however, is populated by all kinds of people, all with good and bad intentions.

Because of the security challenges online, users should stop viewing the Internet as some sort of an upscale suburbia, where everything is nice and quiet and all the houses have white picket fences and well-manicured lawns. Anyone can just leave the house with doors unlocked.

Quite the contrary, people should view the Net as some sort of a tough Chicago neighborhood or Smokey Mountain—a slum area in Manila, where people hatch all kinds of malicious and criminal activities. The only way to live sanely and safely in these neighborhoods is to lock doors, put metal bars on windows and make sure the yard is well lit and equipped with security camera. This should be our mindset every time we walk on the streets of the information superhighway.

Microsoft has been living in the suburbia when it comes to offering products to users. Internet Explorer and its email application, Outlook, have been the most vulnerable to hackers, virus, worms and other problems that wreck havoc on millions of computer users worldwide. Although Microsoft has provided fixes to vulnerabilities in its software application such as Service Pack 2 for Windows and other updates for its Internet Explorer browser, many problems still exist. Hackers still find it easy to introduce all kinds of things from worms, viruses, spyware, adware and others that tend to seriously compromise the security of all the files and transactions done on millions of household computers worldwide.

Online journalists, designers and producers need to be aware of the common security pitfalls of running a new site. The following are the most common problems online:

1. Phishing—A copycat of a famous company or organization website, designed to lure unwitting users to give personal or private information. For example, a customer accesses an email account and finds a legitimate looking message from Discover Card or Citibank saying that your password has expired please click on the link below to provide us with your most current information. If you give out your account name, password, or any other pertinent information, then you might just have been a victim of identity theft.

A simple way to detect this type of fraudulent activity is to look at the IP address. Normally, you are being linked to a non-secure site (http site not https) and it's not even a site that carries the exact company address. The best solution however is to contact the company and ask them for verification or go directly to the company site by typing the web address in your browser toolbar.

Phishing could severely impact an online news operation if someone on staff had fallen prey to a fictitious email from the company or Internet service provider and had given out his or her account name and password. This then allows the hacker to go into the site and get or upload all sorts of things.

Phishing is a hybrid form of hacking. Hacking is the most direct form of security assault on a site. This happens when a person or group of people are able to enter the inner operation of your site and do a lot of damage to it. However, there are software available now that can detect and respond to a few pings or attempts to hack the site.

2. Trojan Horse—Just like in the movie Troy, an otherwise impregnable computer can be penetrated by this type of malicious virus, which surreptitiously enters your hard drive while you access the Internet. As soon as it hits your computer, the virus reads through your files, getting passwords, bank account numbers and whatever information that can be of benefit to a third party wanting to steal precious information from you.

These days this type of an assault can be prevented by subscribing to an anti-virus software service such as the ones developed by Symantec and MacAfee.

3. Worm—The best way to deal with worms is to stay away from them, meaning, don't open any email coming from an unfamiliar source. Some users get attracted to catchy message headings such as "I love you" or any other headings about how to enlarge certain parts of your body, or how to look good, or get rich. They click on the message and open the attachments and then boom, literally, a can of worms have just been opened. This is a form of virus that can damage your files and can go into your email and start copying and sending the infected files to the accounts in your address book. Before a virus fix can be made available, extensive damage may have already been done to your computer and perhaps to your entire network, which then can cripple or slow your online operation to a crawl.

4. Spyware—Unbeknownst to them, every time people access a website, a file settles on your computer called spyware. This file records your browsing patterns and begins serving up pop-ads the next time you visit the web. If you're fond of going to adult sites, then you might well see a slew of pop up advertising asking you to check out site X.

Spywares are especially found in file sharing programs and some audio or video downloads that come from an unfamiliar source. If you have downloaded music or applications through Grokster or Kazaa, there is a high likelihood that your computer has spyware. If you've downloaded the Paris Hilton sex video, then chances are you might have spyware sitting on your hard drive and doing its work of serving up annoying pop-ups.

Aside from annoyance, spyware poses the greatest threat to privacy because it can remember personal information and passwords accessible to a third party.

Some software can remove spywares, bur sometimes it is very difficult to totally clean up the mess. Despite numerous attempts to disinfect your computer, some spywares are simply difficult to detect and remove. Some people have resorted to the radical solution of totally wiping out all files and start from the beginning, meaning re-load everything from the Windows operation system and other software.

Microsoft is now beginning to offer some solution to this problem. Their programmers have made available a free anti spyware on their site Microsoft.com. Ultimately, Microsoft wants to compete with Symantec and MacAffee in selling not only anti-virus but also anti-spyware software to consumers.

Although at this point, Microsoft's Internet Explorer is still the most vulnerable browser from the onslaught of spyware. Some people and even some universities such as Penn State have started using Mozilla Firefox, a free downloadable browser that offers a lot more security features than Explorer and less vulnerability to spyware and the annoying popups.

How is a cookie different from a spyware? The main difference is a cookie can be easily removed by simply clicking the Internet Options command and then removing them. A spyware, meanwhile, requires another application or software for its removal.

A cookie is like a cached file that sits on your Internet file folders and remembers your browsing patterns while a spyware remembers your files, passwords and sites visited, and then shoots advertising as you browse the Net. A cookie can be filtered or blocked, however, some free sites such as the nytimes.com and washingtonpost.com will not allow you to browse their pages if you refused to accept their cookie.

Ethical Challenges

The main reason why copyright laws are in existence is to protect the creator or artist. It is an incentive for artists to continue producing useful work. With the advent of the Internet, it has proven to be difficult to protect copyright from illegal downloading, copying and other forms of infringement.

Associations representing the movie and music industries, for example, have begun cracking on young people downloading or sharing music and other files on the Internet. Aside from this move, encryption technology has significantly improved in the past few years that it is proving to be very difficult to copy or use many files from music to program applications without getting a license or expressed permission from the owner.

For multimedia producers, the rule of thumb should always be this: when in doubt ask permission. Email or call the rightful owner or source of a music, video file before going ahead to use in it online multimedia presentations.

If it is not for profit, the likelihood of getting the green light to use a file is high. There may be some nominal fee sometimes attached to educational use of digital files.

The courts have lately been ruling in heavy favor of the software companies diluting the so-called "fair use" standards, which allows some latitude for users who want to use copyrighted material for educational purposes. Schools are finding it more expensive to acquire software for students to use in producing multimedia interactive projects. It is hoped that this difficulty will soon ease because innovation and discovery rely heavily on the understanding of what's been done so far. One needs old knowledge to create new ones. This is the main reason why companies should go easy on educational institutions that are using their software to harness and create new ones.

10

Marketing and Promoting Online News

It cannot be emphasized enough the role of marketing and promotion in the success of an online news site. Much like the traditional media, dot.com's are also in the business of selling audiences to advertisers. High visitor traffic translates to ad revenues and, ultimately, to the survival of the online news operation.

Managers of online sites are increasingly becoming sophisticated and effective in their efforts to promote traffic to the sites. Some of the promotion efforts out there are the following:

1. Go and do's. This technique showcases the power of convergence and cross-media promotion. The Readership Institute based at Northwestern University in Evanston, IL found that newspapers can help generate more traffic to their website by mentioning the web address in newspaper articles.

Let's say the *Chicago Tribune* published an article about candidates for the Illinois gubernatorial race. In order to help direct readers to its site, the article should carry a "go and do" note telling readers to go checkout the chicagotribune.com for more information about the candidates and their platforms. The promo, for example, might even tell readers to play an interactive game called "Catch George Ryan If You Can."

Television stations are great at the "go and do" technique. This was effectively used during the tsunami relief coverage of 2004 when stations would asked viewers to checkout their website for information on how to help victims of the disaster, which killed a quarter of a million people in southeast Asia.

2. Transcripts, message boards, chat rooms and blogs. These things are fast becoming a regular element of a newspaper site. Some sites offer a way for people to post messages, comments or feedback about any topic. Users can even chat or create a running blog (commentary) in reaction to the articles just posted.

Bloggers or the so-called citizen journalists are often very helpful in providing leads to a story or giving useful feedback on how to go about covering a hot button issue. Aside from the blogging option, it might also be helpful if transcripts of interviews and unclassified data used in investigative pieces are posted online for the readers to evaluate and to make sense of the importance of the said issue.

3. Take a poll. Online polls, especially the interesting ones, usually generate a lot of active participation from visitors. A good way to promote a news site is to publicize polls on the main page or publish the results in traditional media, e.g. newspapers, radio and television.

4. E-mail this. Being able to send articles to another person's email address is one of the best ways to generate buzz for your site. The New York Times Interactive is very good at this. The site not only provides email and printing options for free, it also publicizes the most e-mailed articles during the day, which make people become more curious to know and read why those articles are hot.

5. Multimedia, video and games. No doubt interactive games, multimedia presentations and videos streaming are some of the main traffic drivers to a site. For example, when student journalists at Lewis University started producing and posting video news packages on their site—lewisflyer.com—traffic to the site immediately doubled. Online readers, especially the young, prefer videos than text and they like to be able to customize and interact with the content of a website. That's why sites with interactive games and animations are almost always a big hit. An example of this was the JibJab site where visitors were able to view and e-mail a musical parody of George W. Bush and John Kerry during the presidential campaign of 2004. The animation called "This Land" was one of the most e-mailed files at the time.

6. Traditional and online ads. No doubt that television, radio, newspapers and magazines are still the most catered to media by a large chunk of the population. So it helps to generate site traffic if you're able to advertise your site in other media, especially the ones being followed by your target audience.

Posting banner ads and links on other popular websites is another effective way of promoting a news site. For example, being able to post banner ads or links to popular search engines such as Yahoo! Or Google, or to large Internet service provider sites such as AOL (America Online) or MSN (Microsoft Network) is almost a sure bet that you are going to get clicking and checking out your site.

7. Go mobile. Along with the rise of cell phone use, come mobile Internet and text messaging. Customizing your site to accommodate cell phone users is certainly something worth undertaking. This will help generate more hits to your site, although the effectiveness of the mobile web use is still being tested and explored at this point.

11

Jobs in the Multimedia Marketplace

People who want to pursue a career in the multimedia marketplace need to be aware of the kind of job requirements—knowledge and skills—needed in order to succeed in this highly specialized field. If one needs to survey job postings and newspaper classifieds, the following job categories emerge as the most common in multimedia:

1. **Graphic Designer/Editor**—Companies expect graphic editors to be well versed with Adobe applications such as Photoshop and Illustrator, and to some extent, In Design and Go Live. A graphic editor is expected to design logos, banners and text for a website. He or she is also expected to edit and prepare quality web-ready pictures and images.

2. **Online Producer**—An ideal multimedia producer is someone who has a background in journalistic reporting. Someone who knows how to sift through information and make judgments about what to include in a multimedia project. He or she is someone who has good writing skills, a nose for news, and a good sense of what the online readers are looking for. This person must display good people skills, because he or she will be dealing with sources, talking with other reporters within the news organization, and perhaps working with other journalists outside the organization.

Furthermore, it would be a big plus if a multimedia producer knows non-linear video editing (Avid or Final cut Pro) and certain web applications such as Macromedia Dreamweaver and Flash, the two most commonly used software in producing lively packages on a news site. Knowing the software means being aware of the possibilities in creating online news packages.

3. **Content Editor**—This position is for someone who has strong copy editing skills. Multimedia projects, much like newspaper articles, need careful editing. Otherwise, if the project is hard to grasp and is riddled with glaring typos and grammatical errors, the credibility of the website will suffer a great deal.

4. **Database Editor**—Every online news outfit needs someone who is good at crunching numbers and assembling them in a way that is easy to understand, such charts and bar graphs. A database editor is expected to be highly skilled in computer-assisted reporting, and adept at using a vast array of search engines and databases such as Lexis-Nexis or ProQuest. Because this job also requires a know how in creating databases, an aspiring database editor must at least know how to use Microsoft's Access and Excel. It would even be a big plus if the person knows MySQL and one of the programming languages like Java or Visual Basic.

5. **Programmer**—If a news site contains a lot of dynamic customizable pages, where visitors plug in data and ask a website to spout results, then the news organization definitely needs a programmer who can do the mind-numbing code work and advise the producer and other people involved in planning a multimedia project of what's technically possible and what's not. Oftentimes, some interactive projects are planned without first considering whether it is technically possible and realistic given the short timeframe and limited resources. A programmer should be highly skilled in code languages such as C++, Visual Basic and/or Java.

6. **Web Designer or Multimedia Producer**—This person puts all the elements together—graphics, text, pictures, audio, video—into one big online or interactive package. He or she works closely with all the people involved in the project and has an input at every stage of the planning process. A web designer or producer must be highly skilled in HTML, Macromedia Dreamweaver, Flash, and Maya or Director or any other 3D application. The designer must also be skilled in non-linear editing (Avid or Final Cut Pro) and knows how to use an audio editing software.

Journalism majors are always worried about not getting a job in their field due to a tight labor market. There are too little jobs, too many graduates. This type of thinking is what the economists call as the "lump of labor" fallacy. People think that jobs are finite and there are no new ones out there for fresh graduates.

There are new types of journalism jobs out there, however, they require a far different set of skills than say the type of journalism jobs available 10 or 20 years ago. With the popularity of the Internet and people getting their news from the Web, newsrooms around the world need a new breed of "backpack journalists." These are professionals who know a whole host of skills from reporting to editing, shooting video, interviewing, editing audio and video files and uploading them on the web, and then writing the story for various media. This is the age of convergence and journalists are required to put on many hats on the job.

News companies are investing a lot of money into multimedia journalism. Many newsrooms are having a hard time coping with the changes in the landscape change. Editors and newsroom managers are having hard time training their old dogs some new tricks. And so this is where you as a new journalism or multimedia person come in to fill in the void.

12

Trends and variables in selling online advertising

Can a dotcom owner make money through Internet advertising? If one were to ask this question during the first three years of the new millennium, the answer would most certainly be no—because the online economic bubble at the time had already burst. The price of Internet stocks plunged to its lowest point and the overall national economy slid into a recession. Hundreds of the once lavishly courted and well-paid web designers, programmers, and computer engineers lost their jobs. And companies who relied on a future promise of an increase in Internet advertising revenues were either in the red or bankrupt.

But try asking the question again this time around and one might just get an optimistic reply. Online ad sellers and buyers are finally seeing the bottoming out of the advertising revenue fall. The Internet Advertising Bureau (IAB), an organization whose members come from different advertising sectors (media, creative, account and research), and charged with setting the standards for online advertising[1], projects online ad spending will grow from $6.3 billion in 2003 to $9.3 billion in 2007.[2] Online businesses indeed are now beginning to see a positive turn in profitability.

Challenges still abound.

The loss in revenues during the downturn forced the fledgling new media industry to take a look at the viability of its advertising process, and try to address the concerns raised by buyers who think that online ad prices are sky high and that there are no clear standards in measuring their effectiveness.

For example, today there is still a wide gulf as to how ads are priced. Websites sell them anywhere from a dollar to $100 dollars CPM (cost per thousand users)[3], with the more popular ones like ESPN.com, Nytimes.com and Yahoo! charging advertisers a CPM rate of $35 dollars. MSN (Microsoft Network) charges $10 for every 1,000 ad viewers.[4] Some site owners think that a more reasonable price would be $14 dollars CPM, because it provides a better return on investment.[5]

Meanwhile, online newspapers such as the Nytimes.com and Wall Street Journal Online would like to see day parting as a main determining factor in pricing (IAB still has not addressed this issue). They consider daytime as the equivalent of TV primetime,[6]because Internet traffic usually peaks during the morning and afternoon hours when people are at work. Majority of the people who access the Net everyday do so from work.

The more prudent step to follow, amidst this lack of uniformity, is to look at what the leading sites such as Google, Yahoo!, America Online (AOL), New York Times Online and others are doing on pricing and placement. The type of advertising content and format they allow on their sites can serve as a helpful guide to those new at this venture. IAB has already made some progress in crafting a common format for selling and measuring online ads. Over the next few years, a more standardized ad pricing packages are expected to emerge.

Pricing and measurement criticisms.

The various ways online ads are sold and measured at this time are primarily to blame for the gap in pricing. For instance, some site owners prefer to use hits rather than click-throughs as a unit for gauging the number of visits to a site. Page impression is also another measurement standard being pushed by a growing number of advertisers due to problems found with both hits and click-throughs. Page impression, or more specifically, ad impression is touted as a better alternative because the counting only begins after all the major elements on the site have loaded.

Critics decry the use of hits as a tool for audit and pricing because user figures turn out to be bloated. Computers tend to count any attempt to view a page as a hit even if it did not download completely or even if the browser was closed midway through the process. Also, all the elements on a web page that loaded separately such as banner ads, Flash animation, audio and video are each counted as a hit. So if one uses hits as pricing unit, expect to hear complaints of overcharging from ad buyers.

Click-throughs quickly emerged as an alternative to the much-maligned hits. But its popularity only lasted up to a point—when users were still curious about seeing online ads and would often click on them. Now visitors are used to seeing flashing or animated banner ads all the time they rarely bother to click. The average click-through rate had sunk from about 10% in late 1990s to one-tenth of one percent recently (.10%)[7], which does not bode well for site owners who want to hike revenues through paid advertising.

Just imagine having an extremely popular site with an average of a million visitors per day, the click-through rate would then only equal to 1,000. If ads are sold at a price of $35 CPM (cost per thousand viewers), then the money to be billed is only $35 dollars. This pricing scheme sounds a lot like a losing proposition to site owners considering it takes a lot of resources and manpower to maintain a high-quality interactive site.

Web sites like ESPN.com and others, however, have found a way to use this formula and then bundle it with some kind of profit-sharing arrangement with businesses who buy advertising on their sites. For example, let's say Amazon.com decides to buy advertising on the ESPN site. ESPN then bills the online bookstore company based on the number of click-throughs, plus it will ask for a cut out of the money derived from any site visitor who bought books as a result of his or her clicking on the ad. This click and "per inquiry" combo could be the most viable option to take at the moment, especially for popular sites with high "page view" rates. This selling strategy provides a reliable source of income while allowing web entrepreneurs to chart through the untested waters of online advertising.

Meanwhile, Google, a very popular search engine, offers a slightly different sponsorship ad package called AdWords. Advertisers can pick several keywords that their target customers might use and then display their ads every time those words are used in the search. For instance, a Napa Valley winery had picked "wine" as one of the keywords. Every time a visitor used the word "wine" in his or her search, the wine ad comes up on the results page. Google then bills the company based on an average cost-per-click (CPC), which is about $1.60; the rate tends to go down as the number of clicks go up.[8]

Limiting factors and advances.

Two key factors—bandwidth and memory—lie the future of online advertising. The limited bandwidth and low memory of computers found in millions of households today still serve as stumbling blocks to reaching the highest potentials of online advertising. The Internet is the quintessential example of media convergence. It is television plus interactive, or mass and interpersonal communication combined, in real time. This medium should be a boon to advertisers because, unlike television, it provides a means to track what visitors do online.

But despite the benefits, millions of lifeless sites and document postings still swamp the World Wide Web (WWW). Only those successful commercial sites such as ESPN.com, MSN.com and others have offered a variety of fresh entertainment and interactive content, which keep visitors hooked.

This static print model has rubbed off on Internet advertising as well. Lifeless, billboard type ad images are found in millions of sites worldwide. Talk about reaching young audiences who are used to multi-tasking, instant gratification and fast-paced MTV editing.

If Gordon Moore's law of computing power is any guide, however, a better future for the industry is in the offing. Founder of Intel Corporation, Moore projected back in 1964 that computing power tends to double every 18 months. This means computer innovations will continue to grow. Powerful computers sold in the future should be able to overcome current limitations in speed and space. They are factors that long held back the production and delivery of effective, attention-grabbing online ads.

Even today, changes in format—such as new high-resolution digital video—are already palpable. Online ads are indeed going in the way of television. They now have audio and video or with animation, and most of them are interactive. Advances in software technology such as Macromedia Flash have allowed the creation of engaging animated ads such as the one produced by Orbitz.com and placed on popular newspaper sites. Streaming audio and video software and high-resolution monitors spurred the proliferation of highly appealing 30-second commercials online such as the ones found on Apple.com. Many other companies such as Pepsi, McDonalds, Lexus and IBM had unveiled 15 or 30-second video commercials on the Web in January 2004. Some of these commercials were launched on sites such as ESPN, Lycos and MSN[9].

Typical formats for online ads.

The following are the most common formats for ads online:

1. Rich Media—It is much like a 30-second TV commercial complete with music and videos streamed through Quicktime or Real Media software. Apple for instance has a collection of rich media movie trailers on

its website. Companies are beginning to invest in this format because, according to Nielsen/NetRatings, more and more households (49.5 million or 38% of all households) have subscribed to fast broadband Internet connection. Also, 94% of the 50 million people who surf the Web at work are also connected to broadband.[10]

2. Superstitials—It is normally an animated ad that pops-up on a screen while in the middle of browsing pages. A static or billboard-type version of this ad is commonly called Interstitials. Washingtonpost.com normally carries Superstitials on its pages. It is a little annoying but one always has the option of closing it.

   a. Pop-under—It looks the same as a superstitial only it appears underneath the page. Visitors get to see it when they minimize or close the page currently viewed. Orbitz.com normally has a lot of pop-unders placed on many newspaper sites like Nytimes.com and Yahoo.com.

      Seven percent of all ad revenues come from pop-ups or pop-unders, according to Nielsen/NetRatings. However, this format might soon become extinct because more and more software companies and Internet Service Providers (ISP) are offering pop-up blockers. MSN, AOL, Yahoo! and Google have begun to distribute pop-up blocking software. Microsoft is also expected to put this as built-in software into the new Windows XP operating system[11].

3. Banner ads—This is one of the earliest and still the number one most common format, comprising about a third of all ads, found in millions of sites worldwide. It is often in the form of an animated 6 inches wide and an inch tall ad (468 x 60 pixels) placed on top of a page. Banner ads are also those small rectangular (button) ads placed on either right or left side of the navigation screen.

   a. Square ads—Newspapers online have begun popularizing this type of ad (4 x 4 inches or 250 x 250 pixels), usually placed in the middle of a news story with the text wrapping around it. Some of the ads are quite catchy, animated and directly linked to the advertised product or service [*see Figure 5*].

4. Skyscrapers—The second most popular and an offshoot of the banner ad, a skyscraper is placed vertically on the right side of a page looking

like a tall building (120 x 600). A visitor sees it while scrolling up and down the page.

Exit spam, enter cell phone and IM (instant messaging)

Before the passage of the federal anti-spam law, which took effect in January 2004, e-mail accounts across the U.S. were being bombarded with about 2 billion unsolicited e-mail or spam everyday.[12] Microsoft and the state of New York have filed lawsuits to known spammers, alleging privacy and other illegal business practices. The prospect of costly fines and time in jail effectively put other would-be spammers on notice. This and the enactment of the anti-spam law have led to a sharp decline in the amount of missives sent daily, which used to fill up e-mail spaces to the brim.

Spam is especially pernicious because it is hard to trace its original sender and it is equally tough, if not impossible, to get out of it. The most common types of spam include diet pills, body part enhancers and other health products, which often carry fictitious names or a collection of undecipherable words such as xxT or zzzy as senders. Purveyors of spam charge hundreds of dollars per thousand e-mails sent. They often use third party servers or unknowing computer users to hide themselves.

Spam's pervasiveness did a lot of damage to the image of legitimate e-mail advertising or so-called "permission-based" email advertising. This type of ad has become rare and used only by credit card, airline, catalog, retail stores and other companies whose customers had signed on to be part of a listserv. A visible link is found at the bottom of a page for a member to opt out of the registry, if so desired. The Internet Advertising Bureau (IAB.net) website provides a useful guidelines called "Email Marketing Pledge," which emphasizes consumer consent on all e-mail marketing. It is designed to increase the accountability and credibility of e-mail ads and to "curb unsolicited commercial email." One of its important provisions states that "Commercial email must not be sent to an individual's e-mail address unless one of the following situations exists: I. There is an existing **business relationship**; or II. **Prior informed consent** of the individual has been obtained."[13]

Despite its current lack of popularity, advertisers should not readily discount the effectiveness of "permission-based" email ads. Doubleclick (doubleclick.com), an Internet research firm, conducted a study of e-mail users in the U.S. The study found that 88% of the e-mail ads sent were successfully delivered, 27% were opened and 8.3% had been clicked. Out of the 8.3% click-through rate, 46% indicated that they had "purchased online or offline some time after clicking through an e-mail." They are also more than likely to click and buy a product based on the following reasons: (a) If the ad comes from a known brand; (b) if it carries a special offer and an interesting message; (c) and if the product is relevant to their needs. The Doubleclick e-mail study, however, noted that a great majority (89%) of the respondents still worry about spam. E-mail users (65%) have the tendency to delete "permission-based" ads without reading them or to unsubscribe (24%).[14]

With e-mail advertising moving at a snail's pace, a new type of advertising delivered through wireless or "texting" and instant messaging is on the rise. However, at this point, because the technology is just beginning to reach critical mass, it is still unclear how cell phone carriers structure their rates or how Internet service providers charge companies for ads on their instant messaging software. Yahoo Instant Messaging and AOL Instant Messaging (AIM) are two of the most popular applications techies have quickly adopted, especially those 13 to 29 years old. Instant messaging has now become part of the dynamic of multi-tasking among the youth. And advertising has found its way there too. For example, it is not unusual for Yahoo! Instant Messaging to offer animated advertising or video on a screen before the start of any chat sessions.

Selling online visitors to advertisers

The Internet is now a $300 billion-dollar plus industry and rising. It is the ultimate realization of Marshall McLuhan's global village, where a person can communicate with virtually anyone in the world, instantaneously. Just imagine a Filipino family in Chicago chatting with relatives in the Philippines and viewing video images of each other using a small Webcam connected to Yahoo! Messenger. Imagine a bunch of Certified Public Accounts (CPA) in India preparing tax returns of Americans and sending them back via the H & R Block company website. These worldwide exchanges of data and communications, once thought unimaginable, now seem routine.

There are about 400 million Internet users worldwide and nearly half of them live in the United States.[15] Forty percent of U.S. households have access to the Internet. Of the 130 million who browse the Net everyday, 60% of them access the Internet at work and 40% at home. U.S. Internet users spend an average of 3 hours online per day. Out of this amount, about half an hour is spent chatting and half an hour is spent on e-mail and about 2 hours is spent on browsing the World Wide Web.

The high amount of time spent by Americans on the Internet surely translates into sources of revenue for advertising. It is projected that in 2005 alone, companies will invest in online advertising to the tune of $7 billion dollars. Sales of advertising are expected to grow exponentially beyond that.

Any successful career move towards online advertising sales requires a person to be familiar with user data or audience profiles. And it helps too if one knows the principles and strategies in traditional media selling well because they still very much apply online. Anyone who attended Dr. Greg Pitts' class lecture on media sales would recall an oft-repeated drill line: "the media are in the business of selling audiences to advertisers."

Web service providers can help dotcom owners dig up a wealth of data about activities on a site. Embedded cookies and other software penetrating a person's computer while surfing the Net can provide a helpful clue as to his or her patterns and preferences online. Big time sellers and advertisers alike go to Nielsen/Net Ratings (nielsen-netratings.com), Forrester Research (forrester.com), Jupiter Research (jup.com) and other Internet research firms for the most up-to-date information about Internet use here and around the world.

The worldwide factor

Global reach and instantaneous exchange are two of the most amazing by-products of the Internet revolution. While this sounds like a potential goldmine to site owners who want to make a killing through advertising, it does not readily translate into available dollars. Why aren't advertisers going in droves to buy ads in the most popular of sites? The reality is advertisers, for the most part, have already made up their minds as to whom they want to reach. They know the demographic and lifestyle profiles of their target audience. This behooves dotcom owners to have a clearly defined niche market. All content and promotional efforts are always geared towards driving target users to visit the site frequently.

However, knowing its worldwide reach, some dotcom owners have been successful in expanding their base of online users and have used them as a means to attract more advertising dollars. For example, ESPN.com has expanded its content to include translations and running commentary of games in Mandarin. This occurred because ESPN found that 20% of its site visitors come from

China, a country very much in the minds of U.S. investors these days due to its booming economy, not to mention a billion and a half consumers.

In essence, a successful attempt to sell advertising online requires the possession of a reliable data and a complete knowledge of user profiles. Any attempt to generate more dollars through advertising requires expanding the defined user base by tapping into the global market. To do this, a dotcom enterprise must be able to address several challenges and hurdles, not the least of which language and cultural.

References

Interactive Advertising Bureau (January 2002). *Interactive Audience Measurement and Advertising Campaign Reporting and Audit Guidelines.* [URL] http://www.iab.net/standards/measure_guide.pdf

Greenspan, Robyn (July 15, 2003). *'U.S. Online Ad Growth Underway,"* Cyberatlas [URL] http://cyberatlas.internet.com/markets/advertising/articles/0,1323,5941_2234931,00.html

Angwin, Julia (Spetember 10, 2002). "Web Ads Hit Rock Bottom: Some Are Free," *Wall Street Journal,* pg. B1.

Tedeschi, Bob (2004, Jan. 19) "Television commercials come to the Web," *The New York Times Online.* [URL]: http://www.nytimes.com/2004/01/19/technology/19ecom.html

Paladini, Michael (October 11, 1999). "Make good impressions by good impressions." *Marketing News,* vol. 33, Iss. 21, pg. 17.

Taylor, Catherine (June 30, 2003). "New Media; Old Media; Internet advertising has taken a 180-degree turn." *Brandweek.*

See Paladini, M., p.17

Google Adwords. "Pricing and Billing." [URL] http://adwords.google.com/select/pricing.html

See Tedeschi, B.

See Tedeschi, B. and *see also* Nielsen-Netratings (2004). [URL] http://www.nielsen-netratings.com

Hansell, Saul (2004, Jan. 19). "As consumers revolt, a rush to block pop-up online ads." *The New York Times Online.* [URL] http://www.nytimes.com/2004/01/19/technology/19popup.html?8dpc

"Taking on Junk E-mail" (September 13, 2002). *The New York Times Online.* [URL] http://www.nytimes.com/2002/09/13/opinion/13FRI12.html?pagewanted=printposition=top

Internet Advertising Bureau (2004). *Email Marketing Pledge,* [URL] http://www.iab.net/standards/iab_email_pledge.asp

DoubleClick (October 2003) "*2003 consumer Email Study,*" [URL] http://www.doubleclick.net/us/knowledge

Nielsen-Netratings (2004). "Global Internet Index: Average Usage," [URL] http://www.nielsen-netratings.com

Endnotes

1. Interactive Advertising Bureau (January 2002). *Interactive Audience Measurement and Advertising Campaign Reporting and Audit Guidelines.* [URL] http://www.iab.net/standards/measure_guide.pdf
2. Greenspan, Robyn (July 15, 2003). *'U.S. Online Ad Growth Underway,"* Cyberatlas [URL] http://cyberatlas.internet.com/markets/advertising/articles/0,1323,5941_2234931,00.html
3. Angwin, Julia (Spetember 10, 2002). "Web Ads Hit Rock Bottom: Some Are Free," *Wall Street Journal,* pg. B1.
4. Tedeschi, Bob (2004, Jan. 19) "Television commercials come to the Web," *The New York Times Online.* [URL]: http://www.nytimes.com/2004/01/19/technology/19ecom.html
5. Paladini, Michael (October 11, 1999). "Make good impressions by good impressions." *Marketing News,* vol. 33, Iss. 21, pg. 17.
6. Taylor, Catherine (June 30, 2003). "New Media; Old Media; Internet advertising has taken a 180-degree turn." *Brandweek.*
7. *See* Paladini, M., p.17
8. Google Adwords. "Pricing and Billing." [URL] http://adwords.google.com/select/pricing.html
9. *See* Tedeschi, B.
10. *See* Tedeschi, B. and *see also* Nielsen-Netratings (2004). [URL] http://www.nielsen-netratings.com
11. Hansell, Saul (2004, Jan. 19). "As consumers revolt, a rush to block pop-up online ads." *The New York Times Online.* [URL] http://www.nytimes.com/2004/01/19/technology/19popup.html?8dpc
12. "Taking on Junk E-mail" (September 13, 2002). *The New York Times Online.* [URL] http://www.nytimes.com/2002/09/13/opinion/13FRI12.html?pagewanted=printposition=top

13. Internet Advertising Bureau (2004). *Email Marketing Pledge,* [URL] http://www.iab.net/standards/iab_email_pledge.asp

14. DoubleClick (October 2003) "*2003 consumer Email Study,*" [URL] http://www.doubleclick.net/us/knowledge

15. Nielsen-Netratings (2004). "Global Internet Index: Average Usage," [URL] http://www.nielsen-netratings.com

978-0-595-39708-2
0-595-39708-5

Printed in the United States
87097LV00004BA/259/A